Learn

Eureka Math®
Grade 4
Modules 6 & 7

Published by Great Minds®.

Copyright © 2018 Great Minds®.

Printed in the U.S.A.
This book may be purchased from the publisher at eureka-math.org.
10   9   8   7   6   5   4   3

ISBN 978-1-64054-069-9

G4-M6-M7-L-05.2018

Learn ◆ Practice ◆ Succeed

Eureka Math® student materials for *A Story of Units®* (K–5) are available in the *Learn, Practice, Succeed* trio. This series supports differentiation and remediation while keeping student materials organized and accessible. Educators will find that the *Learn, Practice,* and *Succeed* series also offers coherent—and therefore, more effective—resources for Response to Intervention (RTI), extra practice, and summer learning.

Learn

Eureka Math Learn serves as a student's in-class companion where they show their thinking, share what they know, and watch their knowledge build every day. *Learn* assembles the daily classwork—Application Problems, Exit Tickets, Problem Sets, templates—in an easily stored and navigated volume.

Practice

Each *Eureka Math* lesson begins with a series of energetic, joyous fluency activities, including those found in *Eureka Math Practice.* Students who are fluent in their math facts can master more material more deeply. With *Practice,* students build competence in newly acquired skills and reinforce previous learning in preparation for the next lesson.

Together, *Learn* and *Practice* provide all the print materials students will use for their core math instruction.

Succeed

Eureka Math Succeed enables students to work individually toward mastery. These additional problem sets align lesson by lesson with classroom instruction, making them ideal for use as homework or extra practice. Each problem set is accompanied by a Homework Helper, a set of worked examples that illustrate how to solve similar problems.

Teachers and tutors can use *Succeed* books from prior grade levels as curriculum-consistent tools for filling gaps in foundational knowledge. Students will thrive and progress more quickly as familiar models facilitate connections to their current grade-level content.

Students, families, and educators:

Thank you for being part of the *Eureka Math*® community, where we celebrate the joy, wonder, and thrill of mathematics.

In the *Eureka Math* classroom, new learning is activated through rich experiences and dialogue. The *Learn* book puts in each student's hands the prompts and problem sequences they need to express and consolidate their learning in class.

What is in the Learn book?

Application Problems: Problem solving in a real-world context is a daily part of *Eureka Math.* Students build confidence and perseverance as they apply their knowledge in new and varied situations. The curriculum encourages students to use the RDW process—Read the problem, Draw to make sense of the problem, and Write an equation and a solution. Teachers facilitate as students share their work and explain their solution strategies to one another.

Problem Sets: A carefully sequenced Problem Set provides an in-class opportunity for independent work, with multiple entry points for differentiation. Teachers can use the Preparation and Customization process to select "Must Do" problems for each student. Some students will complete more problems than others; what is important is that all students have a 10-minute period to immediately exercise what they've learned, with light support from their teacher.

Students bring the Problem Set with them to the culminating point of each lesson: the Student Debrief. Here, students reflect with their peers and their teacher, articulating and consolidating what they wondered, noticed, and learned that day.

Exit Tickets: Students show their teacher what they know through their work on the daily Exit Ticket. This check for understanding provides the teacher with valuable real-time evidence of the efficacy of that day's instruction, giving critical insight into where to focus next.

Templates: From time to time, the Application Problem, Problem Set, or other classroom activity requires that students have their own copy of a picture, reusable model, or data set. Each of these templates is provided with the first lesson that requires it.

Where can I learn more about Eureka Math *resources?*

The Great Minds® team is committed to supporting students, families, and educators with an ever-growing library of resources, available at eureka-math.org. The website also offers inspiring stories of success in the *Eureka Math* community. Share your insights and accomplishments with fellow users by becoming a *Eureka Math* Champion.

Best wishes for a year filled with aha moments!

Jill Diniz

Jill Diniz
Director of Mathematics
Great Minds

The Read–Draw–Write Process

The *Eureka Math* curriculum supports students as they problem-solve by using a simple, repeatable process introduced by the teacher. The Read–Draw–Write (RDW) process calls for students to

1. Read the problem.

2. Draw and label.

3. Write an equation.

4. Write a word sentence (statement).

Educators are encouraged to scaffold the process by interjecting questions such as

- What do you see?

- Can you draw something?

- What conclusions can you make from your drawing?

The more students participate in reasoning through problems with this systematic, open approach, the more they internalize the thought process and apply it instinctively for years to come.

Contents

Module 6: Decimal Fractions

Module 7: Exploring Measurement with Multiplication

Grade 4
Module 6

Name _____   Date _____

1. Shade the first 7 units of the tape diagram. Count by tenths to label the number line using a fraction and a decimal for each point. Circle the decimal that represents the shaded part.

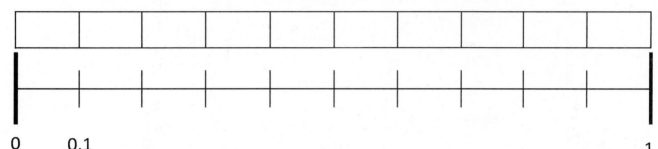

0   0.1   ___   ___   ___   ___   ___   ___   ___   ___   1

         $\frac{1}{10}$

2. Write the total amount of water in fraction form and decimal form. Shade the last bottle to show the correct amount.

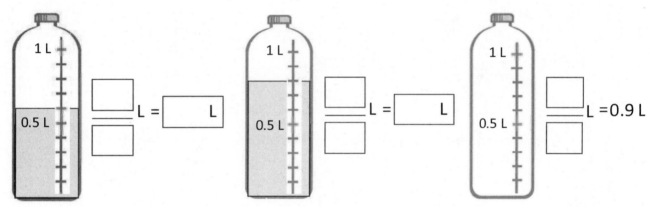

3. Write the total weight of the food on each scale in fraction form or decimal form.

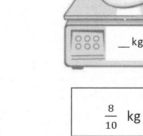

0.4 kg   ___ kg

| kg | $\frac{8}{10}$ kg | kg |

EUREKA
MATH®

Lesson 1:   Use metric measurement to model the decomposition of one whole into tenths.

© 2018 Great Minds®. eureka-math.org

3

4. Write the length of the bug in centimeters. (The drawing is not to scale.)

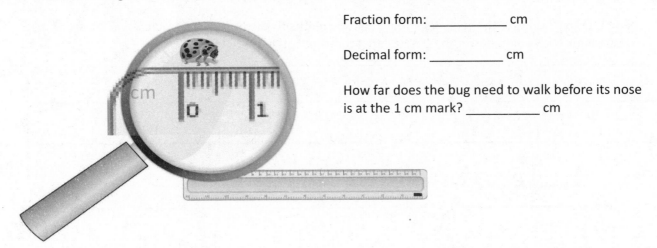

Fraction form: _____ cm

Decimal form: _____ cm

How far does the bug need to walk before its nose is at the 1 cm mark? _____ cm

5. Fill in the blank to make the sentence true in both fraction form and decimal form.

a. $\frac{8}{10}$ cm + _____ cm = 1 cm   0.8 cm + _____ cm = 1.0 cm

b. $\frac{2}{10}$ cm + _____ cm = 1 cm   0.2 cm + _____ cm = 1.0 cm

c. $\frac{6}{10}$ cm + _____ cm = 1 cm   0.6 cm + _____ cm = 1.0 cm

6. Match each amount expressed in unit form to its equivalent fraction and decimal forms.

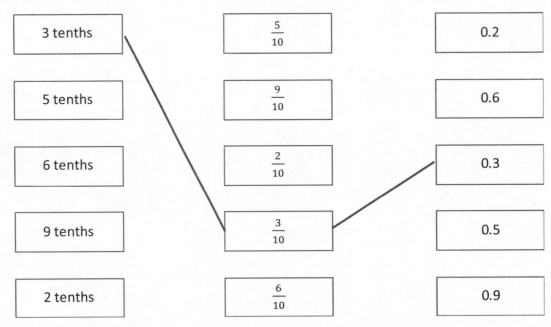

Lesson 1:   Use metric measurement to model the decomposition of one whole into tenths.

© 2018 Great Minds®. eureka-math.org

EUREKA MATH®

Name _____   Date _____

1. Fill in the blank to make the sentence true in both fraction form and decimal form.

a. $\frac{9}{10}$ cm + _____ cm = 1 cm   0.9 cm + _____ cm = 1.0 cm

b. $\frac{4}{10}$ cm + _____ cm = 1 cm   0.4 cm + _____ cm = 1.0 cm

2. Match each amount expressed in unit form to its fraction form and decimal form.

3 tenths		$\frac{5}{10}$		0.8
8 tenths		$\frac{8}{10}$		0.3
5 tenths		$\frac{3}{10}$		0.5

Lesson 1:   Use metric measurement to model the decomposition of one whole into tenths.

Yesterday, Ben's bamboo plant grew 0.5 centimeter. Today it grew another $\frac{8}{10}$ centimeter.

How many centimeters did Ben's bamboo plant grow in 2 days?

_____

_____

_____

Read   **Draw**   **Write**

Lesson 2:   Use metric measurement and area models to represent tenths as
fractions greater than 1 and decimal numbers.

7

Name _____ Date _____

1. For each length given below, draw a line segment to match. Express each measurement as an equivalent mixed number.

   a. 2.6 cm

   b. 3.4 cm

   c. 3.7 cm

   d. 4.2 cm

   e. 2.5 cm

2. Write the following as equivalent decimals. Then, model and rename the number as shown below.

   a. 2 ones and 6 tenths = _____

   $2\frac{6}{10} = 2 + \frac{6}{10} = 2 + 0.6 = 2.6$

Lesson 2: Use metric measurement and area models to represent tenths as fractions greater than 1 and decimal numbers.

9

b. 4 ones and 2 tenths = _____

c. $3\frac{4}{10}$ = _____

d. $2\frac{5}{10}$ = _____

How much more is needed to get to 5? _____

e. $\frac{37}{10}$ = _____

How much more is needed to get to 5? _____

Lesson 2:   Use metric measurement and area models to represent tenths as fractions greater than 1 and decimal numbers.

© 2018 Great Minds®. eureka-math.org

EUREKA
MATH®

Name _____   Date _____

1. For the length given below, draw a line segment to match. Express the measurement as an equivalent mixed number.

   4.8 cm

2. Write the following in decimal form and as a mixed number. Shade the area model to match.

   a.   3 ones and 7 tenths = _____ = _____

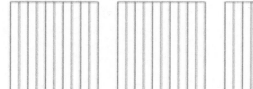  

   b. $\frac{24}{10}$ = _____ = _____

   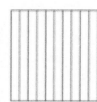 

   How much more is needed to get to 5? _____

tenths area model

Lesson 2: Use metric measurement and area models to represent tenths as fractions greater than 1 and decimal numbers.

13

© 2018 Great Minds®. eureka-math.org

Ed bought 4 pieces of salmon weighing a total of 2 kilograms. One piece weighed $\frac{4}{10}$ kg, and two of the pieces weighed $\frac{5}{10}$ kg each. What was the weight of the fourth piece of salmon?

Read **Draw** **Write**

Lesson 3: Represent mixed numbers with units of tens, ones, and tenths with place value disks, on the number line, and in expanded form.

15

© 2018 Great Minds®. eureka-math.org

Name _____ Date _____

1. Circle groups of tenths to make as many ones as possible.

a. How many tenths in all? 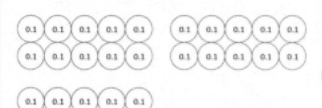 There are _____ tenths.	Write and draw the same number using ones and tenths. Decimal Form: _____ How much more is needed to get to 3? _____
b. How many tenths in all? 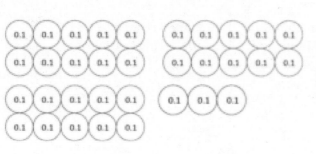 There are _____ tenths.	Write and draw the same number using ones and tenths. Decimal Form: _____ How much more is needed to get to 4? _____

2. Draw disks to represent each number using tens, ones, and tenths. Then, show the expanded form of the number in fraction form and decimal form as shown. The first one has been completed for you.

a. 4 tens 2 ones 6 tenths	b. 1 ten 7 ones 5 tenths
 Fraction Expanded Form $(4 \times 10) + (2 \times 1) + (6 \times \frac{1}{10}) = 42\frac{6}{10}$ Decimal Expanded Form $(4 \times 10) + (2 \times 1) + (6 \times 0.1) = 42.6$	

EUREKA MATH

Lesson 3: Represent mixed numbers with units of tens, ones, and tenths with place value disks, on the number line, and in expanded form.

17

© 2018 Great Minds®. eureka-math.org

c. 2 tens 3 ones 2 tenths	d. 7 tens 4 ones 7 tenths

3. Complete the chart.

Point	Number Line	Decimal Form	Mixed Number (ones and fraction form)	Expanded Form (fraction or decimal form)	How much to get to the next one?
a.			$3\frac{9}{10}$		0.1
b.					
c.				$(7 \times 10) + (4 \times 1) + (7 \times \frac{1}{10})$	
d.			$22\frac{2}{10}$		
e.				$(8 \times 10) + (8 \times 0.1)$	

Lesson 3:   Represent mixed numbers with units of tens, ones, and tenths with place value disks, on the number line, and in expanded form.

EUREKA
MATH®

Name _____   Date _____

1. Circle groups of tenths to make as many ones as possible.

How many tenths in all?	Write and draw the same number using ones and tenths.
0.1 0.1 0.1 0.1 0.1 0.1 0.1 0.1 0.1 0.1 0.1 0.1 0.1 0.1 0.1 0.1 0.1 0.1 There are _____ tenths.	 Decimal Form: _____ How much more is needed to get to 2? _____

2. Complete the chart.

Point	Number Line	Decimal Form	Mixed Number (ones and fraction form)	Expanded Form (fraction or decimal form)	How much to get to the next one?
a.			$12\frac{9}{10}$		
b.		70.7			

Point	Number Line	Decimal Form	Mixed Number (ones and fraction form)	Expanded Form (fraction or decimal form)	How much more is needed to get to the next one?
a.					
b.					
c.					
d.					

tenths on a number line

 **EUREKA MATH**®

Lesson 3: Represent mixed numbers with units of tens, ones, and tenths with place value disks, on the number line, and in expanded form.

© 2018 Great Minds®. eureka-math.org

21

Ali is knitting a scarf that will be 2 meters long. So far, she has knitted $1\frac{2}{10}$ meters.

   a. How many more meters does Ali need to knit to complete the scarf? Write the answer as a fraction and as a decimal.

   b. How many more centimeters does Ali need to knit to complete the scarf?

Read   **Draw**   **Write**

Lesson 4:   Use meters to model the decomposition of one whole into hundredths.
Represent and count hundredths.

© 2018 Great Minds®. eureka-math.org

23

Name _____ Date _____

1. a. What is the length of the shaded part of the meter stick in centimeters?

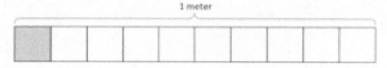

1 meter

    b. What fraction of a meter is 1 centimeter?

    c. In fraction form, express the length of the shaded portion of the meter stick.

1 meter

    d. In decimal form, express the length of the shaded portion of the meter stick.

    e. What fraction of a meter is 10 centimeters?

2. Fill in the blanks.

    a. 1 tenth = _____ hundredths

    b. $\frac{1}{10}$ m = $\frac{}{100}$ m

    c. $\frac{2}{10}$ m = $\frac{20}{}$ m

3. Use the model to add the shaded parts as shown. Write a number bond with the total written in decimal form and the parts written as fractions. The first one has been done for you.

    a.

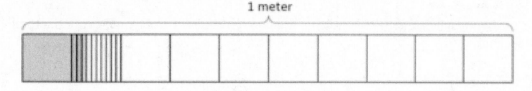

1 meter

0.13

$\frac{1}{10}$ $\frac{3}{100}$

$$\frac{1}{10}\text{m} + \frac{3}{100}\text{m} = \frac{13}{100}\text{m} = 0.13\text{ m}$$

EUREKA MATH

Lesson 4:   Use meters to model the decomposition of one whole into hundredths.
             Represent and count hundredths.

© 2018 Great Minds®. eureka-math.org

25

b.

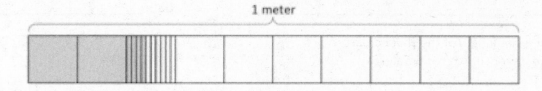

c.

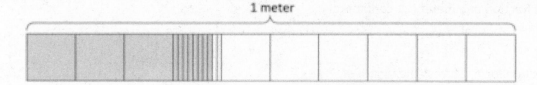

4. On each meter stick, shade in the amount shown. Then, write the equivalent decimal.

a. $\frac{8}{10}$ m

b. $\frac{7}{100}$ m

c. $\frac{19}{100}$ m

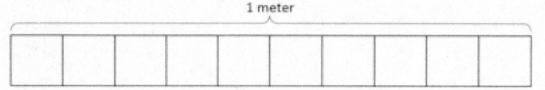

5. Draw a number bond, pulling out the tenths from the hundredths as in Problem 3. Write the total as the equivalent decimal.

a. $\frac{19}{100}$ m

b. $\frac{28}{100}$ m

c. $\frac{77}{100}$

d. $\frac{94}{100}$

Lesson 4:   Use meters to model the decomposition of one whole into hundredths. Represent and count hundredths.

EUREKA
MATH

Name _____   Date _____

1. Shade in the amount shown. Then, write the equivalent decimal.

$\frac{6}{10}$ m

1 meter

2. Draw a number bond, pulling out the tenths from the hundredths. Write the total as the equivalent decimal.

a. $\frac{62}{100}$ m

b. $\frac{27}{100}$

Lesson 4:   Use meters to model the decomposition of one whole into hundredths.
Represent and count hundredths.

© 2018 Great Minds®. eureka-math.org

27

1 meter

1 meter

1 meter

1 meter

1 meter

tape diagram in tenths

Lesson 4:   Use meters to model the decomposition of one whole into hundredths.
             Represent and count hundredths.   29

© 2018 Great Minds®. eureka-math.org

The perimeter of a square measures 0.48 m. What is the measure of each side length in centimeters?

_____

_____

_____

_____

Read   **Draw**   **Write**

Name _____   Date _____

1. Find the equivalent fraction using multiplication or division. Shade the area models to show the equivalency. Record it as a decimal.

   a. $\dfrac{3\times}{10\times} = \dfrac{}{100}$

   b. $\dfrac{50\div}{100\div} = \dfrac{}{10}$

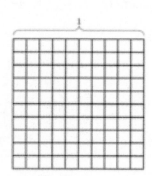
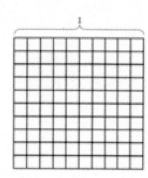

2. Complete the number sentences. Shade the equivalent amount on the area model, drawing horizontal lines to make hundredths.

   a. 37 hundredths = _____ tenths + _____ hundredths

       Fraction form: _____

       Decimal form: _____

   b. 75 hundredths = _____ tenths + _____ hundredths

       Fraction form: _____

       Decimal form: _____

3. Circle hundredths to compose as many tenths as you can. Complete the number sentences. Represent each with a number bond as shown.

   a.

       _____ hundredths = _____ tenth + _____ hundredths

EUREKA MATH

Lesson 5:   Model the equivalence of tenths and hundredths using the area model and place value disks.

33

© 2018 Great Minds®. eureka-math.org

b.

$\bigcirc$ 0.01 $\bigcirc$ 0.01 $\bigcirc$ 0.01 $\bigcirc$ 0.01 $\bigcirc$ 0.01 $\bigcirc$ 0.01 $\bigcirc$ 0.01 $\bigcirc$ 0.01 $\bigcirc$ 0.01 $\bigcirc$ 0.01

$\bigcirc$ 0.01 $\bigcirc$ 0.01 $\bigcirc$ 0.01 $\bigcirc$ 0.01 $\bigcirc$ 0.01 $\bigcirc$ 0.01 $\bigcirc$ 0.01

$\bigcirc$ 0.01 $\bigcirc$ 0.01 $\bigcirc$ 0.01 $\bigcirc$ 0.01 $\bigcirc$ 0.01

$\bigcirc$ 0.01 $\bigcirc$ 0.01 $\bigcirc$ 0.01 $\bigcirc$ 0.01 $\bigcirc$ 0.01   hundredths = _____ tenths + _____ hundredths

4. Use both tenths and hundredths place value disks to represent each number. Write the equivalent number in decimal, fraction, and unit form.

a. $\frac{3}{100}$ = 0. _____ _____ hundredths	b. $\frac{15}{100}$ = 0. _____ _____ tenth _____ hundredths
c. ——— = 0.72 _____ hundredths	d. ——— = 0.80 _____ tenths
e. ——— = 0. _____ 7 tenths 2 hundredths	f. ——— = 0. _____ 80 hundredths

Lesson 5:   Model the equivalence of tenths and hundredths using the area model and place value disks.

© 2018 Great Minds®. eureka-math.org

EUREKA MATH

Name _____   Date _____

Use both tenths and hundredths place value disks to represent each fraction. Write the equivalent decimal, and fill in the blanks to represent each in unit form.

1. $\frac{7}{100}$ = 0. ____

____ hundredths

2. $\frac{34}{100}$ = 0. ____

____ tenths ____ hundredths

Lesson 5:   Model the equivalence of tenths and hundredths using the area model and place value disks.

35

EUREKA MATH®

© 2018 Great Minds®. eureka-math.org

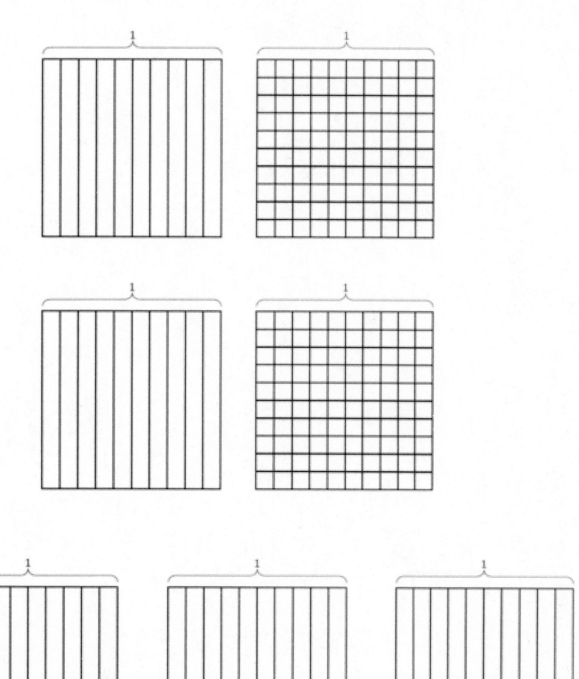

- tenths and hundredths area model

Lesson 5: Model the equivalence of tenths and hundredths using the area model and place value disks.

© 2018 Great Minds®. eureka-math.org

37

The table shows the perimeter of four rectangles.

a. Which rectangle has the smallest perimeter?

Rectangle	Perimeter
A	54 cm
B	$\frac{69}{100}$ m
C	54 m
D	0.8 m

b. The perimeter of Rectangle C is how many meters less than a kilometer?

Read   **Draw**   **Write**

Lesson 6:   Use the area model and number line to represent mixed numbers with
units of ones, tenths, and hundredths in fraction and decimal forms.

39

c. Compare the perimeters of Rectangles B and D. Which rectangle has the greater perimeter? How much greater?

Read   **Draw**   **Write**

Lesson 6:   Use the area model and number line to represent mixed numbers with units of ones, tenths, and hundredths in fraction and decimal forms.

EUREKA MATH

Name _____ Date _____

1. Shade the area models to represent the number, drawing horizontal lines to make hundredths as needed. Locate the corresponding point on the number line. Label with a point, and record the mixed number as a decimal.

   a. $1\frac{15}{100}$ = ___ . _____

   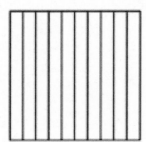 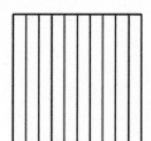 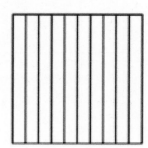

   b. $2\frac{47}{100}$ = ___ . _____

   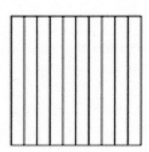 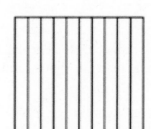 

2. Estimate to locate the points on the number lines.

   a. $2\frac{95}{100}$ b. $7\frac{52}{100}$

Lesson 6:   Use the area model and number line to represent mixed numbers with
             units of ones, tenths, and hundredths in fraction and decimal forms.

41

© 2018 Great Minds®. eureka-math.org

EUREKA MATH®

3. Write the equivalent fraction and decimal for each of the following numbers.

a. 1 one 2 hundredths	b. 1 one 17 hundredths
c. 2 ones 8 hundredths	d. 2 ones 27 hundredths
e. 4 ones 58 hundredths	f. 7 ones 70 hundredths

4. Draw lines from dot to dot to match the decimal form to both the unit form and fraction form. All unit forms and fractions have at least one match, and some have more than one match.

7 ones 13 hundredths ●

7 ones 3 hundredths ●

7 ones 3 tenths ●

7 tens 3 ones ●

● 7.30 ●

● 7.3 ●

● 7.03 ●

● 7.13 ●

● 73 ●

● $7\frac{3}{100}$

● 73

● $7\frac{13}{100}$

● $7\frac{30}{100}$

Lesson 6: Use the area model and number line to represent mixed numbers with units of ones, tenths, and hundredths in fraction and decimal forms.

EUREKA MATH

Name _____ Date _____

1. Estimate to locate the points on the number lines. Mark the point, and label it as a decimal.

   a. $7\frac{20}{100}$

   7   8

   b. $1\frac{75}{100}$

   1   2

2. Write the equivalent fraction and decimal for each number.

   a. 8 ones 24 hundredths

   b. 2 ones 6 hundredths

EUREKA MATH

Lesson 6:   Use the area model and number line to represent mixed numbers with
units of ones, tenths, and hundredths in fraction and decimal forms.

© 2018 Great Minds®. eureka-math.org

43

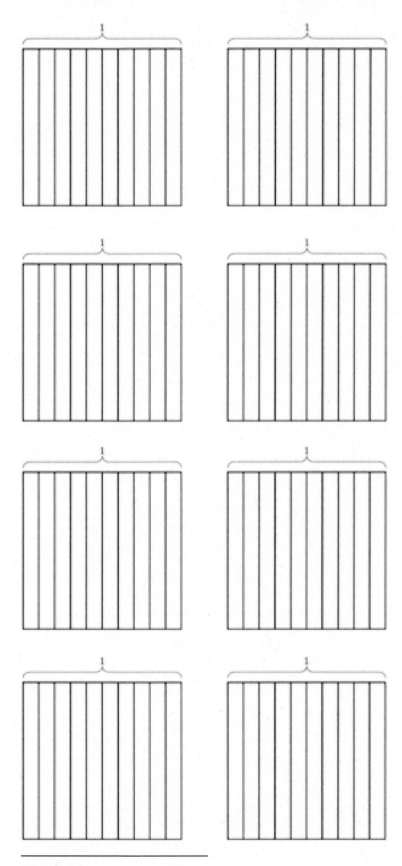

area model

Lesson 6:   Use the area model and number line to represent mixed numbers with units of ones, tenths, and hundredths in fraction and decimal forms.

© 2018 Great Minds®. eureka-math.org

45

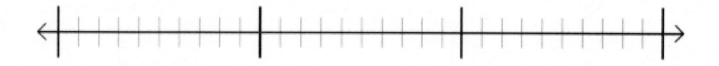

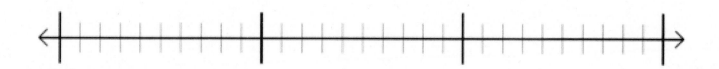

_____

number line

Lesson 6:   Use the area model and number line to represent mixed numbers with units of ones, tenths, and hundredths in fraction and decimal forms.

47

Use pattern blocks to create at least 1 figure with at least 1 line of symmetry. Draw your figure below.

_____

_____

_____

_____

Read   **Draw**   **Write**

Lesson 7:   Model mixed numbers with units of hundreds, tens, ones, tenths, and
hundredths in expanded form and on the place value chart.

49

© 2018 Great Minds®. eureka-math.org

Name _____ Date _____

1. Write a decimal number sentence to identify the total value of the place value disks.

   a.

   2 tens   5 tenths   3 hundredths

   _____ + _____ + _____ = _____

   b.

   5 hundreds   4 hundredths

   _____ + _____ = _____

2. Use the place value chart to answer the following questions. Express the value of the digit in unit form.

hundreds	tens	ones	•	tenths	hundredths
4	1	6		8	3

   a. The digit _____ is in the hundreds place. It has a value of _____.

   b. The digit _____ is in the tens place. It has a value of _____.

   c. The digit _____ is in the tenths place. It has a value of _____.

   d. The digit _____ is in the hundredths place. It has a value of _____.

hundreds	tens	ones	•	tenths	hundredths
5	3	2		1	6

   e. The digit _____ is in the hundreds place. It has a value of _____.

   f. The digit _____ is in the tens place. It has a value of _____.

   g. The digit _____ is in the tenths place. It has a value of _____.

   h. The digit _____ is in the hundredths place. It has a value of _____.

EUREKA MATH

Lesson 7: Model mixed numbers with units of hundreds, tens, ones, tenths, and hundredths in expanded form and on the place value chart.

51

© 2018 Great Minds®. eureka-math.org

3. Write each decimal as an equivalent fraction. Then, write each number in expanded form, using both decimal and fraction notation. The first one has been done for you.

Decimal and Fraction Form	Expanded Form	
	Fraction Notation	Decimal Notation
$15.43 = 15\frac{43}{100}$	$(1\times 10)+(5\times 1)+(4\times\frac{1}{10})+(3\times\frac{1}{100})$ $10\ +\ 5\ +\ \frac{4}{10}\ +\ \frac{3}{100}$	$(1\times 10)+(5\times 1)+(4\times 0.1)+(3\times 0.01)$ $10\ +\ 5\ +\ 0.4\ +\ 0.03$
$21.4 =$ _____		
$38.09 =$ _____		
$50.2 =$ _____		
$301.07 =$ _____		
$620.80 =$ _____		
$800.08 =$ _____		

Lesson 7:   Model mixed numbers with units of hundreds, tens, ones, tenths, and hundredths in expanded form and on the place value chart.

© 2018 Great Minds®. eureka-math.org

EUREKA MATH®

Name _____   Date _____

1. Use the place value chart to answer the following questions. Express the value of the digit in unit form.

hundreds	tens	ones	•	tenths	hundredths
8	2	7		6	4

a   The digit _____ is in the hundreds place. It has a value of _____.

b. The digit _____ is in the tens place. It has a value of _____.

c. The digit _____ is in the tenths place. It has a value of _____.

d. The digit _____ is in the hundredths place. It has a value of _____.

2. Complete the following chart.

Fraction	Expanded Form		Decimal
	Fraction Notation	Decimal Notation	
$422\frac{8}{100}$			
	$(3\times100)+(9\times\frac{1}{10})+(2\times\frac{1}{100})$		

EUREKA MATH®

Lesson 7:   Model mixed numbers with units of hundreds, tens, ones, tenths, and hundredths in expanded form and on the place value chart.

© 2018 Great Minds®. eureka-math.org

53

hundredths	
tenths	
.	
ones	
tens	
hundreds	

place value chart

Lesson 7:   Model mixed numbers with units of hundreds, tens, ones, tenths, and hundredths in expanded form and on the place value chart.

55

© 2018 Great Minds®. eureka-math.org

Jashawn had 5 hundred dollar bills and 6 ten dollar blls in his wallet. Alva had 58 ten dollar bills under her mattress. James had 556 one dollar bills in his piggy bank. They decide to combine their money to buy a computer. Express the total amount of money they have using the following bills:

   a. Hundreds, tens, and ones

   b. Tens and ones

Read   **Draw**   **Write**

Lesson 8:   Use understanding of fraction equivalence to investigate decimal numbers on the place value chart expressed in different units.

c. Ones

Read   **Draw**   **Write**

Lesson 8:   Use understanding of fraction equivalence to investigate decimal
numbers on the place value chart expressed in different units.

© 2018 Great Minds®. eureka-math.org

EUREKA
MATH®

Name _____ Date _____

1. Use the area model to represent $\frac{250}{100}$. Complete the number sentence.

   a. $\frac{250}{100}$ = _____ tenths = _____ ones _____ tenths =__.____

   b. In the space below, explain how you determined your answer to part (a).

2. Draw place value disks to represent the following decompositions:

2 ones = _____ tenths

ones	•	tenths	hundredths

2 tenths = _____ hundredths

ones	•	tenths	hundredths

1 one 3 tenths = _____ tenths

ones	•	tenths	hundredths

2 tenths 3 hundredths = _____ hundredths

ones	•	tenths	hundredths

EUREKA MATH®

Lesson 8: Use understanding of fraction equivalence to investigate decimal numbers on the place value chart expressed in different units.

59

© 2018 Great Minds®. eureka-math.org

3. Decompose the units to represent each number as tenths.

   a. 1 = _____ tenths

   b. 1.7 = _____ tenths

   c. 10.7 = _____ tenths

   b. 2 = _____ tenths

   c. 2.9 = _____ tenths

   d. 20.9 = _____ tenths

4. Decompose the units to represent each number as hundredths.

   a. 1 = _____ hundredths

   b. 1.7 = _____ hundredths

   c. 10.7 = _____ hundredths

   b. 2 = _____ hundredths

   c. 2.9 = _____ hundredths

   d. 20.9 = _____ hundredths

5. Complete the chart. The first one has been done for you.

Decimal	Mixed Number	Tenths	Hundredths
2.1	$2\frac{1}{10}$	21 tenths $\frac{21}{10}$	210 hundredths $\frac{210}{100}$
4.2			
8.4			
10.2			
75.5			

Use understanding of fraction equivalence to investigate decimal numbers on the place value chart expressed in different units.

EUREKA MATH®

Name _____   Date _____

1. a.   Draw place value disks to represent the following decomposition:

3 ones 2 tenths = _____ tenths

ones	·	tenths	hundredths

b.   3 ones 2 tenths = _____ hundredths

2.   Decompose the units.

a.   2.6 = _____ tenths

b.   6.1 = _____ hundredths

Lesson 8:   Use understanding of fraction equivalence to investigate decimal
numbers on the place value chart expressed in different units.

© 2018 Great Minds®. eureka-math.org

61

Tens	Ones		Tenths	Hundredths

area model and place value chart

Lesson 8:   Use understanding of fraction equivalence to investigate decimal
            numbers on the place value chart expressed in different units.

63

© 2018 Great Minds®. eureka-math.org

Kelly's dog weighs 14 kilograms 24 grams. Mary's dog weighs 14 kilograms 205 grams. Hae Jung's dog weighs 4,720 grams.

   a. Order the weight of the dogs in grams from least to greatest.

   b. How much more does the heaviest dog weigh than the lightest dog?

Read   **Draw**   **Write**

    Lesson 9:   Use the place value chart and metric measurement to compare
                    decimals and answer comparison questions.

Name _____ Date _____

1. Express the lengths of the shaded parts in decimal form. Write a sentence that compares the two lengths. Use the expression *shorter than* or *longer than* in your sentence.

   a.
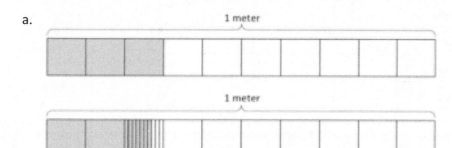

   b.
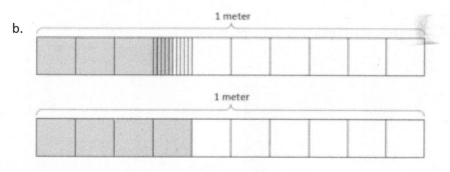

   c. List all four lengths from least to greatest.

2. a. Examine the mass of each item as shown below on the 1-kilogram scales. Put an X over the items that are heavier than the avocado.

Lesson 9: Use the place value chart and metric measurement to compare decimals and answer comparison questions.

67

© 2018 Great Minds®. eureka-math.org

b. Express the mass of each item on the place value chart.

Mass of Fruit (kilograms)

Fruit	ones	.	tenths	hundredths
avocado				
apple				
bananas				
grapes				

c. Complete the statements below using the words *heavier than* or *lighter than* in your statements.

The avocado is _____ the apple.

The bunch of bananas is _____ the bunch of grapes.

3. Record the volume of water in each graduated cylinder on the place value chart below.

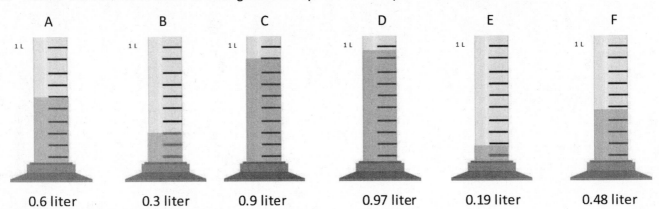

A	B	C	D	E	F
0.6 liter	0.3 liter	0.9 liter	0.97 liter	0.19 liter	0.48 liter

Volume of Water (liters)

Cylinder	ones	.	tenths	hundredths
A				
B				
C				
D				
E				
F				

Compare the values using >, <, or =.

a. 0.9 L _____ 0.6 L

b. 0.48 L _____ 0.6 L

c. 0.3 L _____ 0.19 L

d. Write the volume of water in each graduated cylinder in order from least to greatest.

Lesson 9: Use the place value chart and metric measurement to compare decimals and answer comparison questions.

EUREKA MATH

Name _____   Date _____

1. a.   Doug measures the lengths of three strings and shades tape diagrams to represent the length of each string as show below. Express, in decimal form, the length of each string.

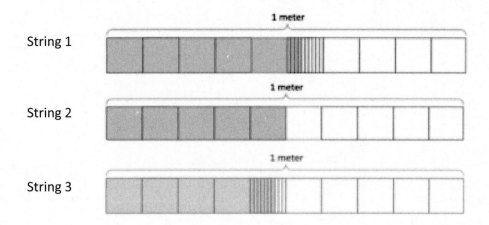

    b.   List the lengths of the strings in order from greatest to least.

2.   Compare the values below using >, <, or =.

    a.   0.8 kg _____ 0.6 kg

    b.   0.36 kg _____ 0.5 kg

    c.   0.4 kg _____ 0.47 kg

EUREKA MATH

Lesson 9:   Use the place value chart and metric measurement to compare decimals and answer comparison questions.

© 2018 Great Minds®. eureka-math.org

69

Mass of Rice Bags (kilograms)

Rice Bag	ones	.	tenths	hundredths
A				
B				
C				
D				

Volume of Liquid (liters)

Cylinder	ones	.	tenths	hundredths
A				
B				
C				
D				

measurement record

Lesson 9: Use the place value chart and metric measurement to compare decimals and answer comparison questions.

71

© 2018 Great Minds®. eureka-math.org

In science class, Emily's 1-liter beaker contains 0.3 liter of water. Ali's beaker contains 0.8 liter of water, and Katie's beaker contains 0.63 liter of water. Who can pour all of her water into Emily's beaker without going over 1 liter, Ali or Katie?

Read   **Draw**   **Write**

Lesson 10:   Use area models and the number line to compare decimal numbers, and record comparisons using <, >, and =.

73

Name _____ Date _____

1. Shade the area models below, decomposing tenths as needed, to represent the pairs of decimal numbers. Fill in the blank with <, >, or = to compare the decimal numbers.

   a. 0.23 _____ 0.4

   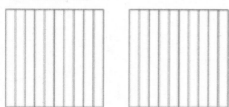

   b. 0.6 _____ 0.38

   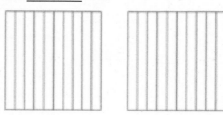

   c. 0.09 _____ 0.9

   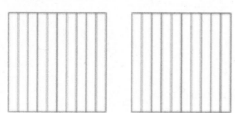

   c. 0.70 _____ 0.7

   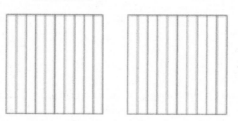

2. Locate and label the points for each of the decimal numbers on the number line. Fill in the blank with <, >, or = to compare the decimal numbers.

   a. 10.03 _____ 10.3

   b. 12.68 _____ 12.8

Lesson 10: Use area models and the number line to compare decimal numbers, and record comparisons using <, >, and =.

75

EUREKA MATH

© 2018 Great Minds®. eureka-math.org

3. Use the symbols <, >, or = to compare.

   a. 3.42 _____ 3.75   b. 4.21 _____ 4.12

   c. 2.15 _____ 3.15   d. 4.04 _____ 6.02

   e. 12.7 _____ 12.70   f. 1.9 _____ 1.21

4. Use the symbols <, >, or = to compare. Use pictures as needed to solve.

   a. 23 tenths _____ 2.3   b. 1.04 _____ 1 one and 4 tenths

   c. 6.07 _____ $6\frac{7}{10}$ d. 0.45 _____ $\frac{45}{10}$

   e. $\frac{127}{100}$ _____ 1.72   f. 6 tenths _____ 66 hundredths

Lesson 10:   Use area models and the number line to compare decimal numbers,
                and record comparisons using <, >, and =.

EUREKA
MATH

Name _____ Date _____

1. Ryan says that 0.6 is less than 0.60 because it has fewer digits. Jessie says that 0.6 is greater than 0.60.
    Who is right? Why? Use the area models below to help explain your answer.

0.6 _____ 0.60

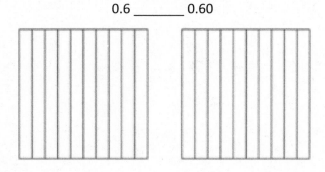

2. Use the symbols <, >, or = to compare.

    a. 3.9 _____ 3.09

    b. 2.4 _____ 2 ones and 4 hundredths

    c. 7.84 _____ 78 tenths and 4 hundredths

EUREKA
MATH®

Lesson 10:   Use area models and the number line to compare decimal numbers,
and record comparisons using <, >, and =.

77

© 2018 Great Minds®. eureka-math.org

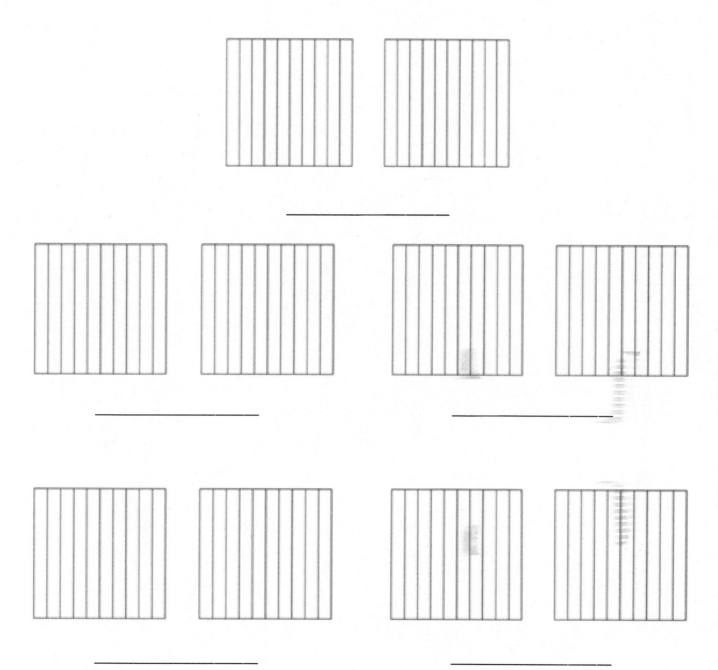

comparing with area models

Lesson 10:   Use area models and the number line to compare decimal numbers,
and record comparisons using <, >, and =.

79

© 2018 Great Minds®. eureka-math.org

While sewing, Kikanza cut 3 strips of colored fabric: a yellow 2.8-foot strip, an orange 2.08-foot strip, and a red 2.25-foot strip.

She put the shortest strip away in a drawer and placed the other 2 strips side by side on a table. Draw a tape diagram comparing the lengths of the strips on the table. Which measurement is longer?

_____

_____

_____

_____

Read   **Draw**   **Write**

© 2018 Great Minds®. eureka-math.org

Name _____   Date _____

1. Plot the following points on the number line.

a.   0.2, $\frac{1}{10}$, 0.33, $\frac{12}{100}$, 0.21, $\frac{32}{100}$

0.1   0.2   0.3   0.4

b.   3.62, 3.7, $3\frac{85}{100}$, $\frac{38}{10}$, $\frac{364}{100}$

3.6   3.7   3.8   3.9

c. $6\frac{3}{10}$, 6.31, $\frac{628}{100}$, $\frac{62}{10}$, 6.43, 6.40

6.2   6.3   6.4   6.5

EUREKA
MATH®

2. Arrange the following numbers in order from greatest to least using decimal form. Use the > symbol between each number.

   a. $\frac{27}{10}$, 2.07, $\frac{27}{100}$, $2\frac{71}{100}$, $\frac{227}{100}$, 2.72

   b. $12\frac{3}{10}$, 13.2, $\frac{134}{100}$, 13.02, $12\frac{20}{100}$

   c. $7\frac{34}{100}$, $7\frac{4}{10}$, $7\frac{3}{10}$, $\frac{750}{100}$, 75, 7.2

3. In the long jump event, Rhonda jumped 1.64 meters. Mary jumped $1\frac{6}{10}$ meters. Kerri jumped $\frac{94}{100}$ meter. Michelle jumped 1.06 meters. Who jumped the farthest?

4. In December, $2\frac{3}{10}$ feet of snow fell. In January, 2.14 feet of snow fell. In February, $2\frac{19}{100}$ feet of snow fell, and in March, $1\frac{1}{10}$ feet of snow fell. During which month did it snow the most? During which month did it snow the least?

Lesson 11: Compare and order mixed numbers in various forms.

EUREKA MATH

Name _____   Date _____

1. Plot the following points on the number line using decimal form.

    1 one and 1 tenth, $\frac{13}{10}$, 1 one and 20 hundredths, $\frac{129}{100}$, 1.11, $\frac{102}{100}$

1.0   1.1   1.2   1.3

2. Arrange the following numbers in order from greatest to least using decimal form. Use the > symbol between each number.

    5.6, $\frac{605}{100}$, 6.15, $6\frac{56}{100}$, $\frac{516}{100}$, 6 ones and 5 tenths

    _____

On Monday, $1\frac{7}{8}$ inches of rain fell. On Tuesday, it rained $\frac{1}{4}$ inch. What was the total rainfall for the two days?

_____

_____

_____

Read   **Draw**   **Write**

Name _____   Date _____

1. Complete the number sentence by expressing each part using hundredths. Model using the place value chart, as shown in part (a).

ones	•	tenths	hundredths
		•	• • • • •
			(• • • • •)
			(• • • • •)

a. 1 tenth + 5 hundredths = _____ hundredths

ones	•	tenths	hundredths

b. 2 tenths + 1 hundredth = _____ hundredths

ones	•	tenths	hundredths

c. 1 tenth + 12 hundredths = _____ hundredths

2. Solve by converting all addends to hundredths before solving.

a. 1 tenth + 3 hundredths = _____ hundredths + 3 hundredths = _____ hundredths

b. 5 tenths + 12 hundredths = _____ hundredths + _____ hundredths = _____ hundredths

c. 7 tenths + 27 hundredths = _____ hundredths + _____ hundredths = _____ hundredths

d. 37 hundredths + 7 tenths = _____ hundredths + _____ hundredths = _____ hundredths

Lesson 12:   Apply understanding of fraction equivalence to add tenths and hundredths.

© 2018 Great Minds®. eureka-math.org

89

3. Find the sum. Convert tenths to hundredths as needed. Write your answer as a decimal.

   a. $\dfrac{2}{10} + \dfrac{8}{100}$

   b. $\dfrac{13}{100} + \dfrac{4}{10}$

   c. $\dfrac{6}{10} + \dfrac{39}{100}$

   d. $\dfrac{70}{100} + \dfrac{3}{10}$

4. Solve. Write your answer as a decimal.

   a. $\dfrac{9}{10} + \dfrac{42}{100}$

   b. $\dfrac{70}{100} + \dfrac{5}{10}$

   c. $\dfrac{68}{100} + \dfrac{8}{10}$

   d. $\dfrac{7}{10} + \dfrac{87}{100}$

5. Beaker A has $\dfrac{63}{100}$ liter of iodine. It is filled the rest of the way with water up to 1 liter. Beaker B has $\dfrac{4}{10}$ liter of iodine. It is filled the rest of the way with water up to 1 liter. If both beakers are emptied into a large beaker, how much iodine does the large beaker contain?

Lesson 12: Apply understanding of fraction equivalence to add tenths and hundredths.

EUREKA MATH

Name _____   Date _____

1. Complete the number sentence by expressing each part using hundredths. Use the place value chart to model.

ones		tenths	hundredths
	•		

1 tenth + 9 hundredths = _____ hundredths

2. Find the sum. Write your answer as a decimal.

$$\frac{4}{10} + \frac{73}{100}$$

EUREKA MATH

Lesson 12:   Apply understanding of fraction equivalence to add tenths and hundredths.

© 2018 Great Minds®. eureka-math.org

91

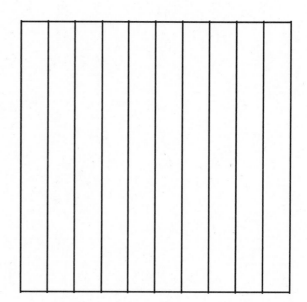

ones		tenths	hundredths
	●		

area model and place value chart

Lesson 12:   Apply understanding of fraction equivalence to add tenths and hundredths.

93

© 2018 Great Minds®. eureka-math.org

Name _____ Date _____

1. Solve. Convert tenths to hundredths before finding the sum. Rewrite the complete number sentence in decimal form. Problems 1(a) and 1(b) are partially completed for you.

a. $2\frac{1}{10} + \frac{3}{100} = 2\frac{10}{100} + \frac{3}{100} =$ _____ $2.1 + 0.03 =$ _____	b. $2\frac{1}{10} + 5\frac{3}{100} = 2\frac{10}{100} + 5\frac{3}{100} =$ _____
c. $3\frac{24}{100} + \frac{7}{10}$	d. $3\frac{24}{100} + 8\frac{7}{10}$

2. Solve. Then, rewrite the complete number sentence in decimal form.

a. $6\frac{9}{10} + 1\frac{10}{100}$	b. $9\frac{9}{10} + 2\frac{45}{100}$
c. $2\frac{4}{10} + 8\frac{90}{100}$	d. $6\frac{37}{100} + 7\frac{7}{10}$

EUREKA MATH

Lesson 13:   Add decimal numbers by converting to fraction form.

95

© 2018 Great Minds®. eureka-math.org

3. Solve by rewriting the expression in fraction form. After solving, rewrite the number sentence in decimal form.

a. 6.4 + 5.3	b. 6.62 + 2.98
c. 2.1 + 0.94	d. 2.1 + 5.94
e. 5.7 + 4.92	f. 5.68 + 4.9
g. 4.8 + 3.27	h. 17.6 + 3.59

Lesson 13:   Add decimal numbers by converting to fraction form.

EUREKA
MATH

Name _____   Date _____

Solve by rewriting the expression in fraction form. After solving, rewrite the number sentence in decimal form.

1. 7.3 + 0.95

2. 8.29 + 5.9

Lesson 13:   Add decimal numbers by converting to fraction form.

© 2018 Great Minds®. eureka-math.org

97

Name _____   Date _____

1. Barrel A contains 2.7 liters of water. Barrel B contains 3.09 liters of water. Together, how much water do the two barrels contain?

2. Alissa ran a distance of 15.8 kilometers one week and 17.34 kilometers the following week. How far did she run in the two weeks?

Lesson 14:   Solve word problems involving the addition of measurements in decimal form.

© 2018 Great Minds®. eureka-math.org

99

3. An apple orchard sold 140.5 kilograms of apples in the morning and 15.85 kilograms more apples in the afternoon than in the morning. How many total kilograms of apples were sold that day?

4. A team of three ran a relay race. The final runner's time was the fastest, measuring 29.2 seconds. The middle runner's time was 1.89 seconds slower than the final runner's. The starting runner's time was 0.9 seconds slower than the middle runner's. What was the team's total time for the race?

Lesson 14:   Solve word problems involving the addition of measurements in decimal form.

EUREKA
MATH

Name _____   Date _____

Elise ran 6.43 kilometers on Saturday and 5.6 kilometers on Sunday. How many total kilometers did she run on Saturday and Sunday?

EUREKA
MATH®

Lesson 14:   Solve word problems involving the addition of measurements in decimal form.

© 2018 Great Minds®. eureka-math.org

101

At the end of the day, Cameron counted the money in his pockets. He counted 7 pennies, 2 dimes, and 2 quarters. Tell the amount of money, in cents, that was in Cameron's pockets.

_____

_____

_____

_____

Read   **Draw**   **Write**

Lesson 15:   Express money amounts given in various forms as decimal numbers.

© 2018 Great Minds®. eureka-math.org

103

Name _____   Date _____

1. 100 pennies = $____.____   100¢ = $\frac{}{100}$ dollar

2. 1 penny = $____.____   1¢ = $\frac{}{100}$ dollar

3. 6 pennies = $____.____   6¢ = $\frac{}{100}$ dollar

4. 10 pennies = $____.____   10¢ = $\frac{}{100}$ dollar

5. 26 pennies = $____.____   26¢ = $\frac{}{100}$ dollar

6. 10 dimes = $____.____   100¢ = $\frac{}{10}$ dollar

7. 1 dime = $____.____   10¢ = $\frac{}{10}$ dollar

8. 3 dimes = $____.____   30¢ = $\frac{}{10}$ dollar

9. 5 dimes = $____.____   50¢ = $\frac{}{10}$ dollar

10. 6 dimes = $____.____   60¢ = $\frac{}{10}$ dollar

11. 4 quarters = $____.____   100¢ = $\frac{}{100}$ dollar

12. 1 quarter = $____.____   25¢ = $\frac{}{100}$ dollar

13. 2 quarters = $____.____   50¢ = $\frac{}{100}$ dollar

14. 3 quarters = $____.____   75¢ = $\frac{}{100}$ dollar

EUREKA MATH

Lesson 15:   Express money amounts given in various forms as decimal numbers.

105

© 2018 Great Minds®. eureka-math.org

Solve. Give the total amount of money in fraction and decimal form.

15. 3 dimes and 8 pennies

16. 8 dimes and 23 pennies

17. 3 quarters 3 dimes and 5 pennies

18. 236 cents is what fraction of a dollar?

Solve. Express the answer as a decimal.

19. 2 dollars 17 pennies + 4 dollars 2 quarters

20. 3 dollars 8 dimes + 1 dollar 2 quarters 5 pennies

21. 9 dollars 9 dimes + 4 dollars 3 quarters 16 pennies

Lesson 15: Express money amounts given in various forms as decimal numbers.

EUREKA MATH

Name _____   Date _____

Solve. Give the total amount of money in fraction and decimal form.

1. 2 quarters and 3 dimes

2. 1 quarter 7 dimes and 23 pennies

Solve. Express the answer as a decimal.

3. 2 dollars 1 quarter 14 pennies + 3 dollars 2 quarters 3 dimes

Lesson 15:   Express money amounts given in various forms as decimal numbers.

107

Name _____   Date _____

Use the RDW process to solve. Write your answer as a decimal.

1. Miguel has 1 dollar bill, 2 dimes, and 7 pennies. John has 2 dollar bills, 3 quarters, and 9 pennies.
    How much money do the two boys have in all?

2. Suilin needs 7 dollars 13 cents to buy a book. In her wallet, she finds 3 dollar bills, 4 dimes, and
    14 pennies. How much more money does Suilin need to buy the book?

3. Vanessa has 6 dimes and 2 pennies. Joachim has 1 dollar, 3 dimes, and 5 pennies. Jimmy has 5 dollars
    and 7 pennies. They want to put their money together to buy a game that costs $8.00. Do they have
    enough money to buy the game? If not, how much more money do they need?

4. A pen costs $2.29. A calculator costs 3 times as much as a pen. How much do a pen and a calculator cost together?

5. Krista has 7 dollars and 32 cents. Malory has 2 dollars and 4 cents. How much money does Krista need to give Malory so that each of them has the same amount of money?

EUREKA
MATH

Name _____ Date _____

Use the RDW process to solve. Write your answer as a decimal.

David's mother told him that he could keep all the money he finds under the sofa cushions in their house. David finds 6 quarters, 4 dimes, and 26 pennies. How much money does David find altogether?

Grade 4
Module 7

Name _____   Date _____

a.

Pounds	Ounces
1	
2	
3	
4	
5	
6	
7	
8	
9	
10	

The rule for converting pounds to ounces is _____.

b.

Yards	Feet
1	
2	
3	
4	
5	
6	
7	
8	
9	
10	

The rule for converting yards to feet is

_____.

c.

Feet	Inches
1	
2	
3	
4	
5	
6	
7	
8	
9	
10	

The rule for converting feet to inches is

_____.

Lesson 1:   Create conversion tables for length, weight, and capacity units using measurement tools, and use the tables to solve problems.

115

Name _____   Date _____

Use RDW to solve Problems 1–3.

1. Evan put a 2-pound weight on one side of the scale. How many 1-ounce weights will he need to put on the other side of the scale to make them equal?

2. Julius put a 3-pound weight on one side of the scale. Abel put 35 1-ounce weights on the other side. How many more 1-ounce weights does Abel need to balance the scale?

3. Mrs. Upton's baby weighs 5 pounds and 4 ounces. How many total ounces does the baby weigh?

4. Complete the following conversion tables, and write the rule under each table.

    a.

Pounds	Ounces
1	
3	
7	
10	
17	

    The rule for converting pounds to ounces is _____.

Lesson 1:   Create conversion tables for length, weight, and capacity units using
             measurement tools, and use the tables to solve problems.

117

© 2018 Great Minds®. eureka-math.org

b.

Feet	Inches
1	
2	
5	
10	
15	

c.

Yards	Feet
1	
2	
4	
10	
14	

The rule for converting feet to inches is

_____.

The rule for converting yards to feet is

_____.

5. Solve.

   a. 3 feet 1 inch = _____ inches

   b. 11 feet 10 inches = _____ inches

   c. 5 yards 1 foot = _____ feet

   d. 12 yards 2 feet = _____ feet

   e. 27 pounds 10 ounces = _____ ounces

   f. 18 yards 9 feet = _____ feet

   g. 14 pounds 5 ounces = _____ ounces

   h. 5 yards 2 feet = _____ inches

6. Answer *true* or *false* for the following statements. If the statement is false, change the right side of the comparison to make it true.

   a. 2 kilograms > 2,600 grams   _____

   b. 12 feet < 140 inches   _____

   c. 10 kilometers = 10,000 meters   _____

Lesson 1: Create conversion tables for length, weight, and capacity units using measurement tools, and use the tables to solve problems.

EUREKA MATH®

Name _____ Date _____

1. Solve.

   a. 8 feet = _____ inches

   b. 4 yards 2 feet = _____ feet

   c. 14 pounds 7 ounces = _____ ounces

2. Answer *true* or *false* for the following statements. If the statement is false, change the right side of the comparison to make it true.

   a. 3 pounds > 60 ounces   _____

   b. 12 yards < 40 feet   _____

EUREKA
MATH®

Lesson 1:   Create conversion tables for length, weight, and capacity units using measurement tools, and use the tables to solve problems.

119

© 2018 Great Minds®. eureka-math.org

Name _____   Date _____

a.

Gallons	Quarts
1	
2	
3	
4	
5	
6	
7	
8	
9	
10	

The rule for converting gallons to quarts is

_____.

b.

Quarts	Pints
1	
2	
3	
4	
5	
6	
7	
8	
9	
10	

The rule for converting quarts to pints is

_____.

c.

Pints	Cups
1	
2	
3	
4	
5	
6	
7	
8	
9	
10	

The rule for converting pints to cups is _____.

d.   1 gallon = ____ pints

1 quart = ____ cups

1 gallon = ____ cups

EUREKA MATH®

Lesson 2:   Create conversion tables for length, weight, and capacity units using measurement tools, and use the tables to solve problems.

121

Name _____   Date _____

Use RDW to solve Problems 1–3.

1. Susie has 3 quarts of milk. How many pints does she have?

2. Kristin has 3 gallons 2 quarts of water. Alana needs the same amount of water but only has 8 quarts. How many more quarts of water does Alana need?

3. Leonard bought 4 liters of orange juice. How many milliliters of juice does he have?

4. Complete the following conversion tables and write the rule under each table.

a.

Gallons	Quarts
1	
3	
5	
10	
13	

The rule for converting gallons to quarts is

_____.

b.

Quarts	Pints
1	
2	
6	
10	
16	

The rule for converting quarts to pints is

_____

Lesson 2: Create conversion tables for length, weight, and capacity units using measurement tools, and use the tables to solve problems.

123

© 2018 Great Minds®. eureka-math.org

5. Solve.

a. 8 gallons 2 quarts = _____ quarts

b. 15 gallons 2 quarts = _____ quarts

c. 8 quarts 2 pints = _____ pints

d. 12 quarts 3 pints = _____ cups

e. 26 gallons 3 quarts = _____ pints

f. 32 gallons 2 quarts = _____ cups

6. Answer true or false for the following statements. If your answer is false, make the statement true.

a. 1 gallon > 4 quarts   _____

b. 5 liters = 5,000 milliliters   _____

c. 15 pints < 1 gallon 1 cup   _____

7. Russell has 5 liters of a certain medicine. If it takes 2 milliliters to make 1 dose, how many doses can he make?

8. Each month, the Moore family drinks 16 gallons of milk and the Siler family goes through 44 quarts of milk. Which family drinks more milk each month?

9. Keith's lemonade stand served lemonade in glasses with a capacity of 1 cup. If he had 9 gallons of lemonade, how many cups could he sell?

Lesson 2:   Create conversion tables for length, weight, and capacity units using measurement tools, and use the tables to solve problems.

© 2018 Great Minds®. eureka-math.org

EUREKA
MATH®

Name _____   Date _____

1. Complete the table.

Quarts	Cups
1	
2	
4	

2. Bonnie's doctor recommended that she drink 2 cups of milk per day. If she buys 3 quarts of milk, will it be enough milk to last 1 week? Explain how you know.

EUREKA MATH®

Lesson 2:   Create conversion tables for length, weight, and capacity units using measurement tools, and use the tables to solve problems.

© 2018 Great Minds®. eureka-math.org

125

Name _____   Date _____

a.

Minutes	Seconds
1	
2	
3	
4	
5	
6	
7	
8	
9	
10	

The rule for converting minutes to seconds is

_____.

b.

Hours	Minutes
1	
2	
3	
4	
5	
6	
7	
8	
9	
10	

The rule for converting hours to minutes is

_____.

c.

Days	Hours
1	
2	
3	
4	
5	
6	
7	
8	
9	
10	

The rule for converting days to hours is

_____.

 EUREKA MATH®

Lesson 3:   Create conversion tables for units of time, and use the tables to solve
problems.

127

© 2018 Great Minds®. eureka-math.org

Name _____ Date _____

Use RDW to solve Problems 1–2.

1. Courtney needs to leave the house by 8:00 a.m. If she wakes up at 6:00 a.m., how many minutes does she have to get ready? Use the number line to show your work.

2. Giuliana's goal was to run a marathon in under 6 hours. What was her goal in minutes?

3. Complete the following conversion tables and write the rule under each table.

a.

Hours	Minutes
1	
3	
6	
10	
15	

The rule for converting hours to minutes and minutes to seconds is

_____.

b.

Days	Hours
1	
2	
5	
7	
10	

The rule for converting days to hours is

_____.

Lesson 3:   Create conversion tables for units of time, and use the tables to solve problems.

129

© 2018 Great Minds®. eureka-math.org

4. Solve.

a. 9 hours 30 minutes = _____ minutes   b. 7 minutes 45 seconds = _____ seconds

c. 9 days 20 hours = _____ hours   d. 22 minutes 27 seconds = _____ seconds

e. 13 days 19 hours = _____ hours   f. 23 hours 5 minutes = _____ minutes

5. Explain how you solved Problem 4(f).

6. How many seconds are in 14 minutes 43 seconds?

7. How many hours are there in 4 weeks 3 days?

Lesson 3:   Create conversion tables for units of time, and use the tables to solve
              problems.

© 2018 Great Minds®. eureka-math.org

EUREKA
MATH

Name _____   Date _____

The astronauts from Apollo 17 completed 3 spacewalks while on the moon for a total duration of 22 hours 4 minutes. How many minutes did the astronauts walk in space?

Lesson 3:   Create conversion tables for units of time, and use the tables to solve
problems.

131

© 2018 Great Minds®. eureka-math.org

Name _____ Date _____

Use RDW to solve the following problems.

1. Beth is allowed 2 hours of TV time each week. Her sister is allowed 2 times as much. How many minutes of TV can Beth's sister watch?

2. Clay weighs 9 times as much as his baby sister. Clay weighs 63 pounds. How much does his baby sister weigh in ounces?

3. Helen has 4 yards of rope. Daniel has 4 times as much rope as Helen. How many more feet of rope does Daniel have compared to Helen?

Lesson 4:   Solve multiplicative comparison word problems using measurement conversion tables.

© 2018 Great Minds®. eureka-math.org

133

4. A dishwasher uses 11 liters of water for each cycle. A washing machine uses 5 times as much water as a dishwasher uses for each load. Combined, how many milliliters of water are used for 1 cycle of each machine?

5. Joyce bought 2 pounds of apples. She bought 3 times as many pounds of potatoes as pounds of apples. The melons she bought were 10 ounces lighter than the total weight of the potatoes. How many ounces did the melons weigh?

Lesson 4: Solve multiplicative comparison word problems using measurement conversion tables.

EUREKA MATH

Name _____   Date _____

Use RDW to solve the following problem.

Brian has a melon that weighs 3 pounds. He cut it into six equal pieces. How many ounces did each piece weigh?

Lesson 4:   Solve multiplicative comparison word problems using measurement
conversion tables.

© 2018 Great Minds®. eureka-math.org

135

Name _____   Date _____

1. a.   Label the rest of the tape diagram below. Solve for the unknown.

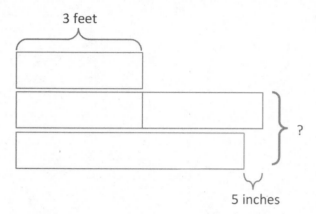

3 feet

?

5 inches

    b.   Write a problem of your own that could be solved using the diagram above.

2.   Create a problem of your own using the diagram below, and solve for the unknown.

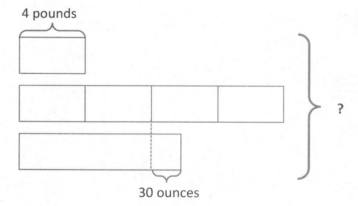

4 pounds

?

30 ounces

Name _____   Date _____

Caitlin ran 1,680 feet on Monday and 2,340 feet on Tuesday. How many yards did she run in those two days?

Classmate:		Problem Number:	
Strategies my classmate used:			
Things my classmate did well:			
Suggestions for improvement:			
Changes I would make to my work based on my classmate's work:			

Classmate:		Problem Number:	
Strategies my classmate used:			
Things my classmate did well:			
Suggestions for improvement:			
Changes I would make to my work based on my classmate's work:			

peer share and critique form

Lesson 5:   Share and critique peer strategies.

Name _____   Date _____

1. Determine the following sums and differences. Show your work.

    a.   3 qt + 1 qt = _____ gal   b.   2 gal 1 qt + 3 qt = _____ gal

    c.   1 gal − 1 qt = _____ qt   d.   5 gal − 1 qt = _____ gal _____ qt

    e.   2 c + 2 c = _____ qt   f.   1 qt 1 pt + 3 pt = _____ qt

    g.   2 qt − 3 pt = _____ pt   h.   5 qt − 3 c = _____ qt _____ c

2. Find the following sums and differences. Show your work.

    a.   6 gal 3 qt + 3 qt = _____ gal _____ qt   b.   10 gal 3 qt + 3 gal 3 qt = _____ gal _____ qt

    c.   9 gal 1 pt − 2 pt = _____ gal _____ pt   d.   7 gal 1 pt − 2 gal 7 pt = _____ gal _____ pt

    e.   16 qt 2 c + 4 c = _____ qt _____ c   f.   6 gal 5 pt + 3 gal 3 pt = _____ gal _____ pt

3. The capacity of a pitcher is 3 quarts. Right now, it contains 1 quart 3 cups of liquid. How much more liquid can the pitcher hold?

4. Dorothy follows the recipe in the table to make her grandma's cherry lemonade.

   a. How much lemonade does the recipe make?

Cherry Lemonade	
Ingredient	**Amount**
Lemon Juice	5 pints
Sugar Syrup	2 cups
Water	1 gallon 1 quart
Cherry Juice	3 quarts

   b. How many more cups of water could Dorothy add to the recipe to make an exact number of gallons of lemonade?

Name _____   Date _____

1. Find the following sums and differences. Show your work.

    a.   7 gal 2 qt + 3 gal 3 qt = _____ gal _____ qt

    b.   9 gal 1 qt − 5 gal 3 qt = _____ gal _____ qt

2. Jason poured 1 gallon 1 quart of water into an empty 2-gallon bucket. How much more water can be added to reach the bucket's 2-gallon capacity?

Samantha is making punch for a class picnic. There are 26 students in her class. Samantha uses 1 gallon 2 quarts of orange juice, 3 quarts of lemonade, and 1 gallon 3 quarts of sparkling water. How much punch did Samantha make? Will there be enough for each student to have two 1-cup servings of punch?

_____

_____

_____

_____

Read   **Draw**   **Write**

Name _____   Date _____

1. Determine the following sums and differences. Show your work.

    a.   1 ft + 2 ft = _____ yd   b.   3 yd 1 ft + 2 ft = _____ yd

    c.   1 yd – 1 ft = _____ ft   d.   8 yd – 1 ft = _____ yd _____ ft

    e.   3 in + 9 in = _____ ft   f.   6 in + 9 in = _____ ft _____ in

    g.   1 ft – 8 in = _____ in   h.   5 ft – 8 in = _____ ft _____ in

2. Find the following sums and differences. Show your work.

    a.   5 yd 2 ft + 2 ft = _____ yd _____ ft   b.   7 yd 2 ft + 2 yd 2 ft = _____ yd _____ ft

    c.   4 yd 1 ft – 2 ft = _____ yd _____ ft   d.   6 yd 1 ft – 2 yd 2 ft = _____ yd _____ ft

    e.   6 ft 9 in + 4 in = _____ ft _____ in   f.   4 ft 4 in + 3 ft 11 in = _____ ft _____ in

    g.   34 ft 4 in – 8 in = _____ ft _____ in   h.   7 ft 1 in – 5 ft 10 in = _____ ft _____ in

3. Matthew is 6 feet 2 inches tall. His little cousin Emma is 3 feet 6 inches tall. How much taller is Matthew than Emma?

4. In gym class, Jared climbed 10 feet 4 inches up a rope. Then, he continued to climb up another 3 feet 9 inches. How high did Jared climb?

5. A quadrilateral has a perimeter of 18 feet 2 inches. The sum of three of the sides is 12 feet 4 inches.

    a. What is the length of the fourth side?

    b. An equilateral triangle has a side length equal to the fourth side of the quadrilateral. What is the perimeter of the triangle?

Lesson 7:   Solve problems involving mixed units of length.

EUREKA
MATH

Name _____   Date _____

Determine the following sums and differences. Show your work.

1.   4 yd 1 ft + 2 ft _____ yd

2.   6 yd – 1 ft = _____ yd _____ ft

3.   4 yd 1 ft + 3 yd 2 ft = _____ yd

4.   8 yd 1 ft – 3 yd 2 ft = _____ yd _____ ft

A sign next to the roller coaster says a person must be 54 inches tall to ride. At his last doctor's appointment, Hever was 4 feet 4 inches tall. He has grown 3 inches since then.

a. Is Hever tall enough to ride the roller coaster? By how many inches does he make or miss the minimum height?

b. Hever's father is 6 feet 3 inches tall. How much taller than the minimum height is his father?

Read   **Draw**   **Write**

Name _____　　Date _____

1. Determine the following sums and differences. Show your work.

   a. 7 oz + 9 oz = _____ lb

   b. 1 lb 5 oz + 11 oz = _____ lb

   c. 1 lb – 13 oz = _____ oz

   d. 12 lb – 4 oz = _____ lb _____ oz

   e. 3 lb 9 oz + 9 oz = _____ lb _____ oz

   f. 30 lb 9 oz + 9 lb 9 oz _____ lb _____ oz

   g. 25 lb 2 oz – 14 oz = _____ lb _____ oz

   h. 125 lb 2 oz – 12 lb 3 oz = _____ lb _____ oz

2. The total weight of Sarah and Amanda's full backpacks is 27 pounds. Sarah's backpack weighs 15 pounds 9 ounces. How much does Amanda's backpack weigh?

3. In Emma's supply box, a pencil weighs 3 ounces. Her scissors weigh 3 ounces more than the pencil, and a bottle of glue weighs three times as much as the scissors. How much does the bottle of glue weigh in pounds and ounces?

4. Use the information in the chart about Jodi's school supplies to answer the following questions:

    a. On Mondays, Jodi packs only her laptop and supply case into her backpack. How much does her full backpack weigh?

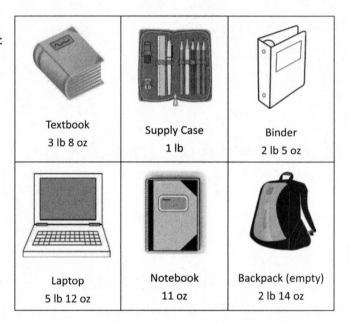

Textbook	Supply Case	Binder
3 lb 8 oz	1 lb	2 lb 5 oz
Laptop	Notebook	Backpack (empty)
5 lb 12 oz	11 oz	2 lb 14 oz

    b. On Tuesdays, Jodi brings her laptop, supply case, two notebooks, and two textbooks in her backpack. On Fridays, Jodi only packs her binder and supply case. How much less does Jodi's full backpack weigh on Friday than it does on Tuesday?

EUREKA
MATH®

Name _____ Date _____

Determine the following sums and differences. Show your work.

1.   4 lb 6 oz + 10 oz = _____ lb _____ oz

2.   12 lb 4 oz + 3 lb 14 oz = _____ lb _____ oz

3.   5 lb 4 oz – 12 oz = _____ lb _____ oz

4.   20 lb 5 oz – 13 lb 7 oz = _____ lb _____ oz

Name _____   Date _____

1. Determine the following sums and differences. Show your work.

    a. 23 min + 37 min = _____ hr

    b. 1 hr 11 min + 49 min = _____ hr

    c. 1 hr – 12 min = _____ min

    d. 4 hr – 12 min = _____ hr _____ min

    e. 22 sec + 38 sec = _____ min

    f. 3 min – 45 sec = _____ min _____ sec

2. Find the following sums and differences. Show your work.

    a. 3 hr 45 min + 25 min = _____ hr _____ min

    b. 2 hr 45 min + 6 hr 25 min = _____ hr _____ min

    c. 3 hr 7 min – 42 min = _____ hr _____ min

    d. 5 hr 7 min – 2 hr 13 min = _____ hr _____ min

    e. 5 min 40 sec + 27 sec = _____ min _____ sec

    f. 22 min 48 sec – 5 min 58 sec = _____ min _____ sec

3. At the cup-stacking competition, the first place finishing time was 1 minute 52 seconds. That was 31 seconds faster than the second place finisher. What was the second place time?

4. Jackeline and Raychel have 5 hours to watch three movies that last 1 hour 22 minutes, 2 hours 12 minutes, and 1 hour 57 minutes, respectively.

   a. Do the girls have enough time to watch all three movies? Explain why or why not.

   b. If Jackeline and Raychel decide to watch only the two longest movies and take a 30-minute break in between, how much of their 5 hours will they have left over?

Lesson 9: Solve problems involving mixed units of time.

EUREKA MATH

Name _____   Date _____

Find the following sums and differences. Show your work.

1. 2 hr 25 min + 25 min = _____ hr _____ min

2. 4 hr 45 min + 2 hr 35 min = _____ hr _____ min

3. 11 hr 6 min – 32 min = _____ hr _____ min

4. 8 hr 9 min – 6 hr 42 min = _____ hr _____ min

Name _____ Date _____

Use RDW to solve the following problems.

1. Paula's time swimming in the Ironman Triathlon was 1 hour 25 minutes. Her time biking was 5 hours longer than her swimming time. She ran for 4 hours 50 minutes. How long did it take her to complete all three parts of the race?

2. Nolan put 7 gallons 3 quarts of gas into his car on Monday and twice as much on Saturday. What was the total amount of gas put into the car on both days?

3. One pumpkin weighs 7 pounds 12 ounces. A second pumpkin weighs 10 pounds 4 ounces. A third pumpkin weighs 2 pounds 9 ounces more than the second pumpkin. What is the total weight of all three pumpkins?

4. Mr. Lane is 6 feet 4 inches tall. His daughter, Mary, is 3 feet 8 inches shorter than her father. His son is 9 inches taller than Mary. How many inches taller is Mr. Lane than his son?

Lesson 10:   Solve multi-step measurement word problems.

Name _____   Date _____

Use RDW to solve the following problem.

Hadley spent 1 hour and 20 minutes completing her math homework, 45 minutes completing her social studies homework, and 30 minutes studying her spelling words. How much time did Hadley spend on homework and studying?

Name _____   Date _____

Use RDW to solve the following problems.

1. Lauren ran a marathon and finished 1 hour 15 minutes after Amy, who had a time of 2 hours 20 minutes. Cassie finished 35 minutes after Lauren. How long did it take Cassie to run the marathon?

2. Chef Joe has 8 lb 4 oz of ground beef in his freezer. This is $\frac{1}{3}$ of the amount needed to make the number of burgers he planned for a party. If he uses 4 oz of beef for each burger, how many burgers is he planning to make?

3. Sarah read for 1 hour 17 minutes each day for 6 days. If she took 3 minutes to read each page, how many pages did she read in 6 days?

4. Grades 3, 4, and 5 have their annual field day together. Each grade level is given 16 gallons of water. If there are a total of 350 students, will there be enough water for each student to have 2 cups?

Name _____ Date _____

Use RDW to solve the following problem.

Judy spent 1 hour 15 minutes less than Sandy exercising last week. Sandy spent 50 minutes less than Mary, who spent 3 hours at the gym. How long did Judy spend exercising?

A rectangular tile has a width of 1 foot 6 inches and length of 2 feet. What is the perimeter of the tile?

_____

_____

_____

_____

Read   **Draw**   **Write**

Lesson 12:   Use measurement tools to convert mixed number measurements to smaller units.

© 2018 Great Minds®. eureka-math.org

171

Name _____ Date _____

1. Draw a tape diagram to show 1 yard divided into 3 equal parts.

   a. $\frac{1}{3}$ yd = _____ ft

   b. $\frac{2}{3}$ yd = _____ ft

   c. $\frac{3}{3}$ yd = _____ ft

2. Draw a tape diagram to show $2\frac{2}{3}$ yards = 8 feet.

3. Draw a tape diagram to show $\frac{3}{4}$ gallon = 3 quarts.

4. Draw a tape diagram to show $3\frac{3}{4}$ gallons = 15 quarts.

5. Solve the problems using whatever tool works best for you.

   a. $\frac{1}{12}$ ft = _____ in

   b. $\frac{}{12}$ ft = $\frac{1}{2}$ ft = _____ in

   c. $\frac{}{12}$ ft = $\frac{1}{4}$ ft = _____ in

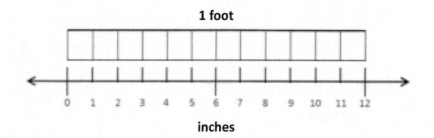

1 foot

inches

EUREKA MATH

Lesson 12: Use measurement tools to convert mixed number measurements to smaller units.

173

© 2018 Great Minds®. eureka-math.org

d. $\dfrac{}{12}$ ft = $\dfrac{3}{4}$ ft = _____ in

e. $\dfrac{}{12}$ ft = $\dfrac{1}{3}$ ft = _____ in

f. $\dfrac{}{12}$ ft = $\dfrac{2}{3}$ ft = _____ in

6. Solve.

a. $1\dfrac{1}{3}$ yd = _____ ft	b. $4\dfrac{2}{3}$ yd = _____ ft
c. $2\dfrac{1}{2}$ gal = _____ qt	d. $7\dfrac{3}{4}$ gal = _____ qt
e. $1\dfrac{1}{2}$ ft = _____ in	f. $6\dfrac{1}{2}$ ft = _____ in
g. $1\dfrac{1}{4}$ ft = _____ in	h. $6\dfrac{1}{4}$ ft = _____ in

Lesson 12: Use measurement tools to convert mixed number measurements to smaller units.

EUREKA MATH

Name _____   Date _____

1. Solve the problems using whatever tool works best for you.

   a. $\frac{}{12}$ ft = $\frac{1}{2}$ ft = _____ in

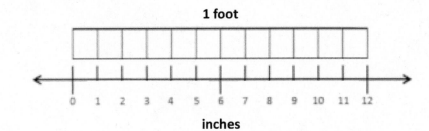

1 foot

inches

   b. $\frac{}{12}$ ft = $\frac{3}{4}$ ft = _____ in

2. Solve.

   a. $1\frac{1}{3}$ yd = _____ ft

   b. $5\frac{3}{4}$ gal = _____ qt

Lesson 12:   Use measurement tools to convert mixed number measurements to smaller units.

175

EUREKA
MATH®

© 2018 Great Minds®. eureka-math.org

Micah used $3\frac{3}{4}$ gallons of paint to paint his bathroom. He used 3 times as much paint to paint his bedroom. How many quarts of paint did it take to paint his bedroom?

_____

_____

_____

_____

Read   **Draw**   **Write**

Lesson 13:   Use measurement tools to convert mixed number measurements to smaller units.

177

© 2018 Great Minds®. eureka-math.org

Name _____   Date _____

1. Solve.

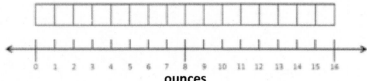
ounces

a. $\frac{1}{16}$ pound = _____ ounce

b. $\frac{}{16}$ pound = $\frac{1}{2}$ pound = _____ ounces

c. $\frac{}{16}$ pound = $\frac{1}{4}$ pound = _____ ounces

d. $\frac{}{16}$ pound = $\frac{3}{4}$ pound = _____ ounces

e. $\frac{}{16}$ pound = $\frac{1}{8}$ pound = _____ ounces

f. $\frac{}{16}$ pound = $\frac{3}{8}$ pound = _____ ounces

2. Draw a tape diagram to show $2\frac{1}{2}$ pounds = 40 ounces.

3.

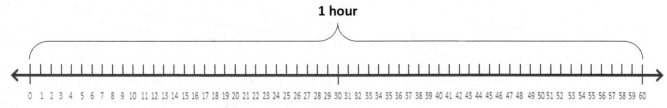
1 hour

0 1 2 3 4 5 6 7 8 9 10 11 12 13 14 15 16 17 18 19 20 21 22 23 24 25 26 27 28 29 30 31 32 33 34 35 36 37 38 39 40 41 42 43 44 45 46 47 48 49 50 51 52 53 54 55 56 57 58 59 60

minutes

a. $\frac{1}{60}$ hour = _____ minute

b. $\frac{}{60}$ hour = $\frac{1}{2}$ hour = _____ minutes

c. $\frac{}{60}$ hour = $\frac{1}{4}$ hour = _____ minutes

4. Draw a tape diagram to show that $1\frac{1}{2}$ hours = 90 minutes.

Lesson 13:   Use measurement tools to convert mixed number measurements to smaller units.

179

© 2018 Great Minds®. eureka-math.org

5. Solve.

a. $1\frac{1}{8}$ pounds = _____ ounces	b. $3\frac{3}{8}$ pounds = _____ ounces
c. $5\frac{3}{4}$ lb = _____ oz	d. $5\frac{1}{2}$ lb = _____ oz
e. $1\frac{1}{4}$ hours = _____ minutes	f. $3\frac{1}{2}$ hours = _____ minutes
g. $2\frac{1}{4}$ hr = _____ min	h. $5\frac{1}{2}$ hr = _____ min
i. $3\frac{1}{3}$ yards = _____ feet	j. $7\frac{2}{3}$ yd = _____ ft
k. $4\frac{1}{2}$ gallons = _____ quarts	l. $6\frac{3}{4}$ gal = _____ qt
m. $5\frac{3}{4}$ feet = _____ inches	n. $8\frac{1}{3}$ ft = _____ in

Lesson 13:   Use measurement tools to convert mixed number measurements to smaller units.

EUREKA MATH

Name _____ Date _____

1. Draw a tape diagram to show that $4\frac{3}{4}$ gallons = 19 quarts.

2. Solve.

a. $1\frac{1}{4}$ pounds = _____ ounces	b. $2\frac{3}{4}$ hr = _____ min
c. $5\frac{1}{2}$ feet = _____ inches	d. $3\frac{5}{6}$ ft = _____ in

EUREKA MATH

Lesson 13: Use measurement tools to convert mixed number measurements to smaller units.

© 2018 Great Minds®. eureka-math.org

181

Name _____ Date _____

Use RDW to solve the following problems.

1. A cartoon lasts $\frac{1}{2}$ hour. A movie is 6 times as long as the cartoon. How many minutes does it take to watch both the cartoon and the movie?

2. A large bench is $7\frac{1}{6}$ feet long. It is 17 inches longer than a shorter bench. How many inches long is the shorter bench?

3. The first container holds 4 gallons 2 quarts of juice. The second container can hold $1\frac{3}{4}$ gallons more than the first container. Altogether, how much juice can the two containers hold?

Lesson 14: Solve multi-step word problems involving converting mixed number measurements to a single unit.

183

4. A girl's height is $3\frac{1}{3}$ feet. A giraffe's height is 3 times that of the girl's. How many inches taller is the giraffe than the girl?

5. Five ounces of pretzels are put into each bag. How many bags can be made from $22\frac{3}{4}$ pounds of pretzels?

6. Twenty servings of pancakes require 15 ounces of pancake mix.

   a. How much pancake mix is needed for 120 servings?

   b. Extension: The mix is bought in $2\frac{1}{2}$-pound bags. How many bags will be needed to make 120 servings?

Lesson 14:   Solve multi-step word problems involving converting mixed number measurements to a single unit.

EUREKA MATH

Name _____   Date _____

Use RDW to solve the following problem.

It took Gigi 1 hour and 20 minutes to complete a bicycle race. It took Johnny twice as long because he got a flat tire. How many minutes did it take Johnny to finish the race?

Lesson 14:   Solve multi-step word problems involving converting mixed number measurements to a single unit.

185

© 2018 Great Minds®. eureka-math.org

Emma's rectangular bedroom is 11 ft long and 12 ft wide. Draw and label a diagram of Emma's bedroom. How many square feet of carpet does Emma need to cover her bedroom floor?

_____

_____

_____

_____

Read   **Draw**   **Write**

Name _____   Date _____

1. Emma's rectangular bedroom is 11 ft long and 12 ft wide with an attached closet that is 4 ft by 5 ft. How many square feet of carpet does Emma need to cover both the bedroom and closet?

2. To save money, Emma is no longer going to carpet her closet. In addition, she wants one 3 ft by 6 ft corner of her bedroom to be wood floor. How many square feet of carpet will she need for the bedroom now?

Lesson 15:   Create and determine the area of composite figures.

189

3. Find the area of the figure pictured to the right.

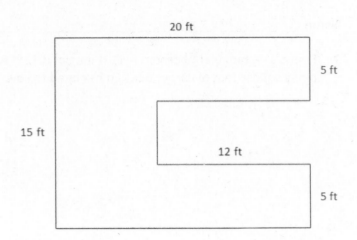

4. Label the sides of the figure below with measurements that make sense. Find the area of the figure.

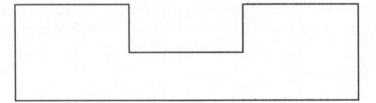

190   Lesson 15:   Create and determine the area of composite figures.

EUREKA
MATH

5. Peterkin Park has a square fountain with a walkway around it. The fountain measures 12 feet on each side. The walkway is $3\frac{1}{2}$ feet wide. Find the area of the walkway.

6. If 1 bag of gravel covers 9 square feet, how many bags of gravel will be needed to cover the entire walkway around the fountain in Peterkin Park?

Lesson 15:   Create and determine the area of composite figures.

191

© 2018 Great Minds®. eureka-math.org

Name _____   Date _____

In the table below are topics that you learned in Grade 4 and that were used in today's lesson.

Choose 1 topic, and describe how you were successful in using it today.

2-digit by 2-digit multiplication	Area formula	Division of 3-digit number by 1-digit number
Subtraction of multi-digit numbers	Addition of multi-digit numbers	Solving multi-step word problems

Lesson 15:   Create and determine the area of composite figures.

193

© 2018 Great Minds®. eureka-math.org

Name _____   Date _____

Work with your partner to create each floor plan on a separate piece of paper, as described below.

You should use a protractor and a ruler to create each floor plan and be sure each rectangle you create has two sets of parallel lines and four right angles.

Be sure to label each part of your model with the correct measurement.

1. The bedroom in Samantha's dollhouse is a rectangle 26 centimeters long and 15 centimeters wide. It has a rectangular bed that is 9 centimeters long and 6 centimeters wide. The two dressers in the room are each 2 centimeters wide. One measures 7 centimeters long, and the other measures 4 centimeters long. Create a floor plan of the bedroom containing the bed and dressers. Find the area of the open floor space in the bedroom after the furniture is in place.

2. A model of a rectangular pool is 15 centimeters long and 10 centimeters wide. The walkway around the pool is 5 centimeters wider than the pool on each of the four sides. In one section of the walkway, there is a flowerbed that is 3 centimeters by 5 centimeters. Create a diagram of the pool area with the surrounding walkway and flowerbed. Find the area of the open walkway around the pool.

Name _____   Date _____

In the table below are skills that you learned in Grade 4 and that you used to complete today's lesson. These skills were originally introduced in earlier grades, and you will continue to work on them as you go on to later grades. Choose three topics from the chart, and explain how you think you might build on and use them in Grade 5.

Multiply 2-digit by 2-digit numbers	Use the area formula to find the area of composite figures	Create composite figures from a set of specifications
Subtract multi-digit numbers	Add multi-digit numbers	Solve multi-step word problems
Construct parallel and perpendicular lines	Measure and construct 90° angles	Measure in centimeters

Lesson 16:   Create and determine the area of composite figures..

197

© 2018 Great Minds®. eureka-math.org

Name _____ Date _____

1. What are you able to do now in math that you were not able to do at the beginning of Grade 4?

2. Which activities would you like to practice this summer in order to keep fluent or become more fluent?

3. What type of practice would help you build your fluency with these concepts?

Name _____   Date _____

1. Why do you think vocabulary was such an important part of fourth-grade math? How does vocabulary help you in math?

2. Which vocabulary terms do you know well, and which would you like to improve upon?

Credits

Great Minds® has made every effort to obtain permission for the reprinting of all copyrighted material. If any owner of copyrighted material is not acknowledged herein, please contact Great Minds for proper acknowledgment in all future editions and reprints of this module.